Around The World With
FOOD AND SPICES
Rice

Melinda Lilly

Rourke Publishing LLC
Vero Beach, Florida 32964

For Mom

PHOTO CREDITS:
Cover photo by Scott M. Thompson • Photo on title page by Scott M. Thompson • Page 7 In the Rice Fields, 1901, by Angelo Morbelli. Photo courtesy of Sotheby's Picture Library, London • Page 8 Photo by Ariel Javellana, International Rice Research Institute • Page 11 Photo by Ariel Javellana, International Rice Research Institute • Page 12 Stationery Box, 17th century Japan. Los Angeles County Museum of Art, Gift of the 1988 Collectors Committee. Photograph ©2000 Museum Associates/LACMA M.88.83a-b • Page 15 Alexander the Great. Workshop of Andrea del Verrocchio. Gift of Therese K. Straus. Photograph © 2000 Board of Trustees, National Gallery of Art, Washington • Page 17 Types of Javanese Ships. Johannus Theodorus De Bry. © National Maritime Museum, London. • Page 19 Detail from A Boxed Set of Watercolour Albums of Chinese Subjects. Studio of Ting Qua. Photo courtesy of Sotheby's Picture Library, London Page 20 Photo by Scott M. Thompson • Page 23 Detail of A Man and a Child Riding on a Plantation, by Michael Jean Cabazon. Photo courtesy of Sotheby's Picture Library, London. • Page 24 Thomas Jefferson, by Mather Brown. National Portrait Gallery, Smithsonian Institution; Bequest of Charles Francis Adams • Page 27 Photo courtesy of the Alexander P. Anderson family • Page 28 Photo by Ariel Javellana, International Rice Research Institute • Page 30 Photo by Ariel Javellana, International Rice Research Institute

ILLUSTRATIONS:
Artwork on world map and additional illustrations by Patti Rule
Artwork on cover by James Spence

EDITORIAL SERVICES:
Pamela Schroeder

Library of Congress Cataloging-in-Publication Data

Lilly, Melinda
  Rice / Melinda Lilly.
    p. cm. — (Around the world with food and spices)
  Includes index.
  ISBN 1-58952-047-5
    1. Cookery (Rice)—Juvenile literature. 2. Rice—Juvenile literature. [Rice.] I. Title

TX809.R5 L552001
641.3'318—dc21   00-054322

Printed in the USA

Table of Contents

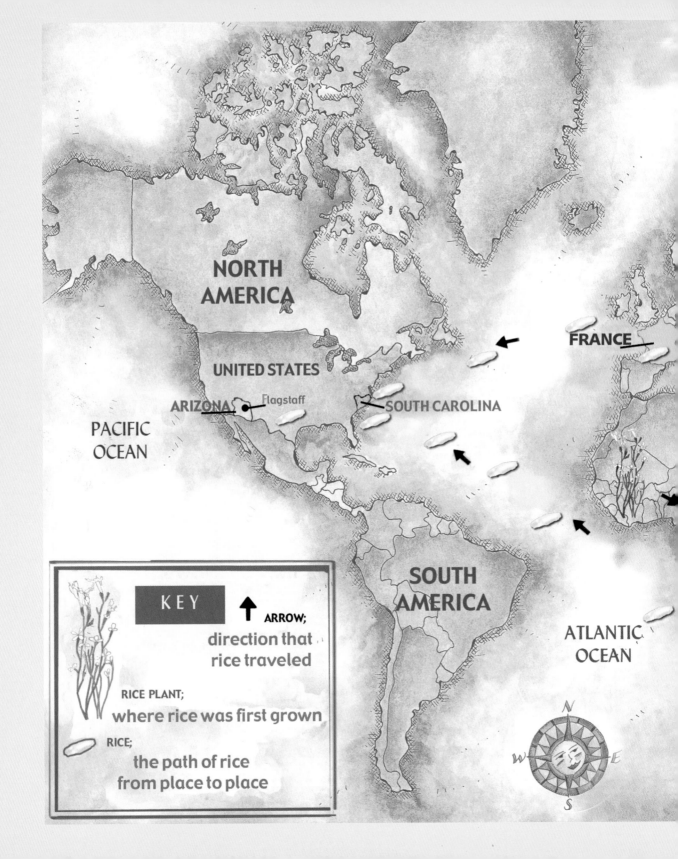

NORTH
AMERICA

UNITED STATES

ARIZONA   Flagstaff

PACIFIC
OCEAN

FRANCE

SOUTH CAROLINA

SOUTH
AMERICA

ATLANTIC
OCEAN

KEY

↑ ARROW;
direction that
rice traveled

RICE PLANT;
where rice was first grown

RICE;
the path of rice
from place to place

N
W E
S

This map shows where rice came from. Follow the rice to see how it reached the United States.

Grain of Life

The mighty dinosaurs stomped through the wild grass, crushing it with each step. Long after the dinosaurs died out, the wild grass won over the world!

Now it is farmed on all the **continents** except Antarctica. It is the main food for more than half the people on Earth. What is this mighty grass? Rice!

Women planting rice

The Rice Bowl

There are more than 120,000 varieties of rice. All these types of rice come from two basic kinds, one from Asia and another from Africa.

Rice, like wheat, is a special sort of grass called **grain**. People in Asia have been gathering and eating the seeds of this grain for at least 6,000 years!

Inside each bud a grain of rice grows.

9

Super Rice Paddy

Rice can be grown on land that gets a lot of rain. However, most rice is grown in a **rice paddy**, a flat field covered by water.

The most amazing rice paddies in the world are in the Philippines. For more than 2,000 years the Ifugao people have grown rice in paddies on steep mountainsides. The paddies' stone walls would stretch half way around the Earth if they were placed in a line!

The Philippine rice paddies are watered by a river that flows from the top of the mountain.

Gift of the Sun or Sea

An old story of Japan tells that the sun goddess brought rice from the heavens as a gift to the islands. Others say that sailors from China were the first to share rice seeds with the Japanese. Most agree that rice has grown in Japan for more than 4,000 years.

In Japan, rice is used to make food, rope, fuel, clothes, hats, bricks, make-up, oil for soap, and more.

Japanese box showing farmers planting rice in Japan

PRæter dictas naues Jauani adhuc genera quatuor in vsu
habent. Vnum grandius, duobus malis & totidem ve-
lis insigne. Secundum minus, quo ferè oram legere, &
merces ex loco in locum traijcere solent. Tertium, sca-
pha piscatoriæ simile: quo tanta concitationẽ & celeritate
vehuntur, vt volitare propè videantur. Unde & voli-
tantes ǹaues dicuntur. Harum carbasa magno numero in Bantam, aliæ
ex arborum folijs, aliæ ex alga conficiuntur. Ad puppim gubernaculum
nullum annectunt, sed vtrinque præfixis remis nauem pro lubitu agunt.

Canoe Trip

More than 2,000 years ago, people from Java set off across the Indian Ocean. They traveled 4,200 miles in canoes that carried rice and other items.

After about two years, they reached Madagascar, an island near Africa. There they sold rice seeds, spices, and other things before heading for home. Rice became so important to the people of Madagascar that it was used as money for hundreds of years.

Canoes and ships of Java, from an old book

Champa Rice

The **emperor** of China was worried. There were many people and only one rice **harvest** per year. Then Emperor Chang Tsung learned that in Champa, Vietnam, there was a variety of rice that would give two harvests per year.

He sent gifts to Champa asking for some of the special rice seed. When the Champa rice arrived in China in the year 1011, people were so happy they wrote poems celebrating it!

A Chinese emperor on a trip

18

The Blessings of Rice

Many countries in Asia have ceremonies to help their rice grow. On the island of Bali, the goddess of life is called the Rice Mother. During the growing season, the farmers gather to honor her.

Long ago, the Chinese emperor would plow his royal rice field and give a handful of dirt to each farmer. They believed that planting the dirt in their own paddies would help their farms.

This Rice Mother doll from Bali is made of palm leaves. Dolls like this one are given to the goddess during holidays.

Rice, Africa, and America

In the 1600s, rice came to South Carolina by way of Africa. The seed came from Madagascar, located off the east coast of Africa. It was West African slaves on American rice **plantations** who knew how to grow rice seed.

Rice has been grown in West Africa for at least 3,500 years. In history there have been many West African kingdoms made rich by their rice farms.

A slave carries rice while a rice plantation owner and his son ride horses.

Thomas Jefferson's Rice

Thomas Jefferson went on a secret **mission** to help the United States in 1787. The new country's farmers needed good rice seeds. Jefferson was in France and thought he could help.

He rode by mule train over the mountainous French **Alps** to the rice fields of Italy. However, no one was allowed to take rice seeds out of Italy. Jefferson hid bags of seed in his pockets and returned home, saving American rice farms!

This portrait of Thomas Jefferson was painted while he was in France helping the United States.

Rice from a Cannon

The crowd held their breath as they watched the eight cannons in the middle of the room. Suddenly the cannons shot puffed rice into the air! Soon everyone was eating treats made from Alexander P. Anderson's new invention—puffed rice. His exploding rice was one of the wonders at the 1904 World's Fair.

Two years before, Anderson had learned how to make rice "pop" by using special cannons and steam. Puffed rice quickly became a popular cereal.

Alexander P. Anderson puts a tube of rice into a special oven. Soon the rice will puff.

Golden Grain

Christian Eijkman wondered why people who ate only white rice got sick. His studies in 1901 helped lead to the discovery of **vitamins**, a part of food needed for good health. Scientists found that white rice didn't have enough vitamins.

In 1999, Ingo Potrykus added a vitamin-making part of the daffodil flower to rice. He hoped this golden rice would keep people healthy by giving them vitamin A.

A boy harvesting rice

Good Grain of Life

Over time, people have changed rice so it could feed more people. In 1998 people ate so much rice that if everyone had eaten the same amount each person would have had 128 pounds!

In Chinese, rice is called the "good grain of life." In Arabic it is often simply called "life." Rice is the mighty grass that changed, and keeps changing, our world.

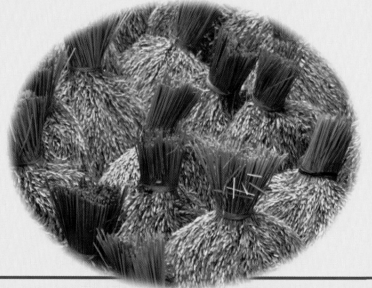

Glossary

alps (ALPS) — high mountains

continents (KON te nents) — the seven main land areas of Earth

emperor (EM per er) — a king who rules many countries

empire (EM pyr) — a nation made up of many countries and ruled by one person

grain (GRAYN) — a small grass seed that can be eaten as food

harvest (HAR vest) — gathering a crop when it is ripe, or a ripe crop of food

mission (MISH en) — a special task given to a person or group

plantations (plan TAY shenz) — large farms, often with workers living on the property

rice paddy (rys PAD ee) — a flat field covered by water where rice is grown

vitamins (VY te minz) — parts of food that help the body to stay healthy

Index

Further Reading

Bracken, Thomas. *Strange Customs of the World*. Chelsea House
    Publishers, 1998.
Robson, Pam. *Rice.* Children's Press, 1998

Websites to Visit

www.riceworld.org
www.riceweb.org

About the Author

Melinda Lilly is the author of several children's books. Some of
her past jobs have included editing children's books, teaching
pre-school, and working as a reporter for *Time* magazine.